Computing and Systems Technology Division 2024

Held at the 2024 AIChE Spring Meeting and 20th Global Congress on Process Safety

New Orleans, Louisiana, USA
24-28 March 2024

ISBN: 979-8-3313-1705-8

Printed from e-media with permission by:

Curran Associates, Inc.
57 Morehouse Lane
Red Hook, NY 12571

AIChE
120 Wall Street, FL 23
New York, NY 10005-4020

Phone: (800) 242-4363
Fax: (203) 775-5177

www.aiche.org

Additional copies of this publication are available from:

Curran Associates, Inc.
57 Morehouse Lane
Red Hook, NY 12571 USA
Phone: 845-758-0400
Fax: 845-758-2633
Email: curran@proceedings.com
Web: www.proceedings.com

TABLE OF CONTENTS

177b Potential for Explainability in Industrially Deployed MPC

Abhijit Bhakte, Abhinav Garg, Rajagopalan Srinivasan

The advent of digitalization in the chemical process industry has brought Artificial Intelligence (AI) into the spotlight, as it offers the potential to optimize operations, increase efficiency, and enhance safety. However, the inherent lack of explainability and transparency in AI systems poses a challenge, making it difficult for non-technical end-users, particularly plant operators, to comprehend the black-box nature of AI-driven decisions. To address this issue, recent years have witnessed the development of novel techniques collectively referred to as Explainable Artificial Intelligence (XAI). XAI methods are pivotal in making AI model results understandable to end-users by providing visual, rule-based, feature-importance, or example-based explanations(Arrieta et al., 2020).

In industry, Model predictive control (MPC) is a widely accepted technology that leverages a process model to obtain optimized control sequences while considering system constraints and process variable interdependencies. Although MPC is not entirely a black box, its complexity often results in decisions that may not align with the intuitive control mechanisms familiar to operators (Lindscheid et al., 2016), thus presenting challenges similar to AI systems. To overcome these challenges, we study the explainability of MPC results generated for operators, thus bridging the gap between operator understanding and MPC decisions.

The Model Predictive Control (MPC) system, in its pursuit of process efficiency, often makes intricate decisions in various scenarios that prove challenging for operators to comprehend. Operating on optimization principles, MPC's methodology is not inherently intuitive for non-technical personnel, including operators. In certain situations, operators may mistakenly perceive MPC as malfunctioning and switch to regulatory / manual controls, even when it functions correctly. This misinterpretation can deviate the process from optimal operation, thereby increasing the operational cost, and significantly increase the workload for operators, particularly in complex industrial settings. The decision to deactivate MPC may arise from the operator's difficulty in grasping the underlying logic of MPC's decision-making. Such issues are exacerbated when operators are not fully acquainted with MPC's optimization objectives or control strategies. For instance, the operator works in the upstream process, and MPC optimizes its objective by considering a few variables from the downstream process, which may not be straightforward for upstream operator to comprehend. Further, such scenarios may offer an opportunity for the designer of MPC to revisit the modeling and optimization framework and make modifications, if required. Therefore, there is a compelling need for MPC systems to maximize transparency, elucidating their operational processes and decision-making mechanisms to enhance mutual understanding between operators and the MPC system. Hence, we use XAI methodologies to interpret the MPC decisions.

XAI techniques can be classified based on how to integrate explanation into the model. Explanations may be an inherent part of a specific machine learning model, known as the intrinsic approach. Notable examples of this approach include decision trees, which rely on hierarchical if-then rules. In contrast, explanations can be introduced as a post-processing step without affecting the model's performance, referred to as the post-hoc approach. Post-hoc methods can be further categorized into model-specific and model-agnostic (Arrieta et al., 2020). Model-specific methods are specially designed for specific machine-learning algorithms such as DNNs. Class Activation Map is one such example that provides a visual explanation for CNNs (Sun et al., 2020). On the other hand, model-agnostic methods work with any type of machine-learning model. The Limit-based explanations for monitoring (LEMON) is a model-agnostic method for explainability because it is independent of internal processing and model representation (Bhakte et al., 2023). The LEMON method uses alarm limits to explain the process monitoring results by building a local model in the vicinity of the process sample. The above taxonomy provides a broad overview of XAI techniques for selecting the appropriate XAI techniques tailored to a specific domain.

In this paper, we underscore the need for explainability of MPC results and demonstrate it through a case study. We utilize a post-hoc model agnostic method for generating explanations as MPC already has an existing model that we want to explain. The results from our study will underline the benefits of explainability in high-stakes decision-making for non-technical end-users, such as operators. In summary, XAI techniques play a vital role in ensuring transparency and comprehensibility, not only in AI systems but also in complex control systems like MPC, ultimately improving decision-making and operational performance in industrial settings.

References

Arrieta, A.B., Díaz-Rodríguez, N., Ser, J.D., Bennetot, A., Tabik, S., Barbado, A., Garcia, S., Gil-Lopez, S., Molina, D., Benjamins, R., Chatila, R., Herrera, F., 2020. Explainable Artificial Intelligence (XAI): Concepts, taxonomies, opportunities and challenges toward responsible AI. Information Fusion 58, 82–115. https://doi.org/10.1016/j.inffus.2019.12.012

Bhakte, A., Chakane, M., Srinivasan, R., 2023. Alarm-based Explanations of Process Monitoring Results from Deep Neural Networks. Computers & Chemical Engineering 179, 108442. https://doi.org/10.1016/j.compchemeng.2023.108442

Lindscheid, C., Bremer, A., Haßkerl, D., Tatulea-Codrean, A., Engell, S., 2016. A Test Environment to Evaluate the Integration of Operators in Nonlinear Model-Predictive Control of Chemical Processes11The research leading to these results has received funding from the European Commission under grant agreement number 291458 (ERC Advanced Investigator Grant MOBOCON). IFAC-PapersOnLine 49, 129–134. https://doi.org/10.1016/j.ifacol.2016.12.202

Sun, K.H., Huh, H., Tama, B.A., Lee, S.Y., Jung, J.H., Lee, S., 2020. Vision-Based Fault Diagnostics Using Explainable Deep Learning With Class Activation Maps. IEEE Access 8, 129169–129179. https://doi.org/10.1109/ACCESS.2020.3009852

141a A Framework for Risk Assessment in the Development of Hydrogen-Based Supply Chains

Austin Johnes, Faisal Khan, Faruque Hasan

With hydrogen production continuing its advance as a viable alternative in the energy sector to traditional energy resources, it is becoming increasingly more important that safety be integrated into the decisions early in the design process of hydrogen supply chains. Typically, the use of metrics such as hazard indices can be incorporated at the process level in the development of safe production designs to assess the relative inherent hazards and/or risk involved. However, risk assessment through hazard indices does have some shortcomings. First, most of these indices were originally designed to be used in traditional chemical processes such as natural gas processing, chemical manufacturing, etc. Hazard indices need to be able to incorporate new innovations as they are developed, such as those that can be used to supplement the development of hydrogen supply chains like energy storage technologies. Secondly, while there has been some development on risk assessment specifically for hydrogen supply chains, these methods lack the scalability that is incorporated with some of the traditional hazard indices. This could potentially lead to either under or overestimates of the risk levels in the supply chain. Finally, as decisions at the supply chain level are typically complex due to uncertainties, geographic specifications, and time variability, a comparative analysis of the relative risks involved might not be sufficient to determine optimal decisions.

Thus, there needs to be a systematic approach that can be implemented at the supply chain level such that the inherent risks are both quantified and fully minimized. As such, we propose a general framework for risk assessment that can be used not only to assess inherent risk relating to the various process units involved in a similar manner to traditional risk/safety indices, but also can be used to incorporate emerging technologies with an estimated risk assessment. To demonstrate the applicability of this framework, we will examine the hydrogen supply chain network. We will demonstrate how this framework can be used to address the risks involved with emerging technologies such as energy storage and how the risk level can be scaled with capacity.

141b Reduced Order CFD Model By an Asymptotic Expansion for Heterogenous Catalysis

Muhammad Uzair Qureshi, Daniel Runge, Georg Brösigke, Christian Merdon, Juergen Fuhrmann, Sebastian Matera, Jens-Uwe Repke

Computational Fluid Dynamics (CFD) analysis of reactive flows over heterogeneous catalysts is a challenging task even for "simple" laminar flows. This involves solving a system of Partial Differential Equations (PDEs) with highly non-linear boundary conditions imposed by the surface chemistry which leads to stiffness and bad conditioning of the overall equation system requiring extensive computational effort. Moreover, the recent development of experimental techniques which provide spatially resolved atomistic scale information of the catalyst under operando conditions makes it possible to derive the underlying catalytic kinetics by solving an inverse problem (which would be only plausible) with a computationally less intensive CFD analysis for such catalytic reactors.

In this study, we present a reduced order approach with a significant lower computational footprint than conventional CFD, which caters for systems with heterogeneous catalytic reactions occurring only at the walls of the reaction chambers where the area of the catalyst is small as compared to the entire flow domain. The approach would be applicable to general reactor geometries and in the scope of 3D printing. Examples of such systems would be the catalyst characterization chambers or catalytic monoliths.

The approach is based on the observation that in many experiments the mixture density and the species transport coefficients are hardly affected by the surface chemistry. The velocity field then decouples from the chemistry and can be obtained from the non-reactive flow problem. This assumption becomes exact in the limit of low chemical reactivity and for excess of one or more species. Focusing on the assumption before, we exploit the fact that in the case of small catalyst limit, the lateral variation of the reactivity on the catalytic surface will be small. So, a leading order asymptotic series expansion for these species results in a problem, which can be solved by three simple steps. First, calculating the velocity field from the non-reactive solution. Second, obtaining the solution of a linear transport problem with constant reactivity. Finally, parametrically feeding of the solution into a small non-linear algebraic problem, which has as many unknowns as species.

The approach has been implemented using VoronoiFVM [1] Finite Volume package programmed in Julia, which guarantees mass conservation. We will demonstrate the approach using the Methanol formation reaction from CO_2 and H_2 and for flow geometries where the zero-reactivity solution for the velocity field is known. Using these, the strengths and limitations of the approach are addressed. The proposed approach is benchmarked against numerical simulations of the fully coupled problem using the CatalyticFoam solver [2,3].

Acknowledgements

We thank the Elsa Neumann Stipendium (Nachwuchsförderungsgesetz NaFöG) and Higher Education Comission Pakistan (ID 57589483) for their financial support for this project.

References:

[1]*Fuhrmann, J.; Abdel, D.; Weidner, J.; Seiler, A.; Farrell, P.; Liero, M. VoronoiFVM.jl - Finite*

volume solver for coupled nonlinear partial differential equations; https://github.com/j-fu/VoronoiFVM.jl; doi:10.5281/zenodo.3529808

[2]*Maestri, M.; Cuoci, A, J. Coupling CFD with detailed microkinetic modeling in heterogeneous catalysis; Chemical Engineering Science; 2013; pp 106-117; doi.org/10.1016/j.ces.2013.03.048*

[3]*OpenFOAM,2011. www.openfoam.org .*

141c Experimental Design for Reaction Rate Model Discrimination

Byanne Malluhi, Radhia Fezai, Mamoun Al-Rawashdeh, Costas Kravaris, Hazem Nounou, Mohamed Nounou

Reaction rate modeling is a crucial tool for understanding and optimizing chemical reactions. Multiple kinetic models can be developed to describe the same reaction, which often poses a challenge for practitioners in selecting the most appropriate model for their studies. In this case, one should perform experiments that will maximize the discrimination capacity. Some of the well-known model discrimination based experimental designs include: Hunter and Reiner's criterion [1], Box and Hill's criterion [2], T-optimality criterion [3], and Akaike weights design criterion AWDC [4]. While various solutions are available for addressing this challenge, emerging optimization tools encourage faster and more efficient methods for discriminating between increasingly more complex models. In this work, we aim to investigate model discrimination and the design-of-experiments capabilities using Bayesian Optimization (BO). BO utilizes a non-parametric Gaussian Process model, which does not have a defined structure and relies on prior knowledge about the behavior of reaction rates. Thus, we will study the usefulness of the proposed strategy in discriminating between nonlinear *parametric* models with as little experiments as possible. To this end, different experimental design procedures for model discrimination will be compared. The proposed methodology will be tested using the Methanol Synthesis Reaction Modeling Case Study [5]. In this case study, twenty kinetic models are proposed and only 27 data points are provided. Each model is a function of four experimental conditions: temperature, partial pressures of CO, H_2, and MeOH.

Keywords:

Experimental design; Bayesian Optimization; Reaction rate modeling; Model Discrimination

References:

[1] W. G. Hunter and A. M. Reiner, "Designs for Discriminating Between Two Rival Models," *Technometrics*, vol. 7, no. 3, pp. 307–323, Aug. 1965, doi: 10.1080/00401706.1965.10490265.

[2] G. E. P. Box and W. J. Hill, "Discrimination among Mechanistic Models," *Technometrics*, vol. 9, no. 1, pp. 57–71, 1967, doi: 10.2307/1266318.

[3] A. C. P. de Leon and A. C. Atkinson, "Optimum Experimental Design for Discriminating Between Two Rival Models in the Presence of Prior Information," *Biometrika*, vol. 78, no. 3, pp. 601–608, 1991, doi: 10.2307/2337029.

[4] L. Cai, S. Kruse, D. Felsmann, C. Thies, K. K. Yalamanchi, and H. Pitsch, "Experimental Design for Discrimination of Chemical Kinetic Models for Oxy-Methane Combustion," *Energy & Fuels*, vol. 31, no. 5, pp. 5533–5542, May 2017, doi: 10.1021/acs.energyfuels.6b03025.

[5] R. J. Berger, J. Hoorn, J. Verstraete, and J. W. Verwijs, "Software functionality assessment for kinetic parameter estimation, model discrimination and design of experiments : The four test cases," *Eurokin*, Apr. 2001, [Online]. Available: http://eurokin.tudelft.nl/?cat=30.

190b Modelling the Feasibility of Variable Electrified Operation: A High Purity Distillation Column Case Study

Isuru A. Udugama, Christoph Bayer, Michael Taube, Brent Young

The transition to renewable electricity grids, and the electrification of industrial processes to replace fossil fuel energy are powerful recent developments that are promising to be potential solutions to achieving significant greenhouse emissions reductions even towards net zero in industrial production processes. Generally, grid electricity is perceived as a stable and reliable resource that can be accessed on demand. However, that is not always the case for large industrial users, and a key challenge that must be overcome is the inherent uncertainty of the electricity supply by renewable sources, as well as in pricing.

In methanol production, the concept of using electrolysis and CO2 capture is being explored as a way of replacing reforming to produce syngas for "green" methanol production. With the omission of the reforming process which supplied energy as steam for the process, there is now a need to electrify the effected downstream unit operations.

Distillation is a key unit operation in chemical operations that consumes a large amount of energy for reboiler duty and holds the key to making on specification AA grade methanol that is used in chemical processes. A key challenge for operating such a distillation column in an electrified operation is not only the inherent uncertainty of availability, but also the pricing of electricity.

In this work we explore the running of a validated digital model of a real distillation plant case study both dynamically to counter uncertainties such as a sudden large reboiler duty drop, as well as through the proposal and the use of the distillation column as a "process battery". To operate the distillation column in such a manner means the traditional process control and operating regimes, that are focused on maintaining steady column operations under all conditions require a major re-think. This work explores, using the case study, the implications of running a high purity distillation column as a "process battery" as a means of countering of extremely volatile reboiler duty availability that is a possibility with electrified operations. A new control philosophy is tested, and its response to a volatile electricity market and/or supply, and conclusions and recommendations for future work are made.

190c Advances in Data-Driven Fault Detection and Identification and Application to a Bench Scale Fischer-Tropsch Process

Nour Basha, Radhia Fezai, Byanne Malluhi, Khaled Dhibi, Gasim Ibrahim, Hanif Choudhury, Nimir Elbashir, Hazem Nounou, Mohamed Nounou

Data-driven modeling is a topic of great interest in current literature, where the core motivation is to find robust black-box models capable of replacing complex mathematical formulas in representing the dynamics of a multivariate system. Black-box models are particularly useful for situations where the mathematical model of the system is unavailable or difficult to accurately estimate. Robust data-driven modeling techniques can then be reliably applied to fault detection and identification applications, where the objective of the former is to identify when the system has deviated outside of the expected normal range of operation. On the other hand, the objective of fault identification is to identify the exact nature of the fault that the system is currently experiencing, if any. Both applications are closely interlinked and highly beneficial for the sake of critical early maintenance and improved safety standards in time-sensitive and/or expensive operations [1-3]. In this work, several linear and non-linear data-driven modeling techniques are going to be applied to real data samples gathered from a bench-scale Fischer-Tropsch process, where the ultimate objective is to compare the accuracy of the models in terms of their robustness in the presence of noise and their ability to detect faults. Linear models include Principal Component Analysis (PCA) and some of its extensions, such as Bayesian PCA and/or Multiscale PCA [4]. Non-linear models include Kernel PCA and Neural Networks [3,5].

Keywords: Fault Detection, Fischer-Tropsch, Data-Driven Modeling.

References

[1] Basha, N., Sheriff, M. Z., Kravaris, C., Nounou, H., Nounou, M., Multiclass data classification using fault detection-based techniques, *Computers & Chemical Engineering*, **136**, 1-11 (2020).

[2] Basha, N., Kravaris, C., Nounou, H., Nounou, M., Bayesian-optimized gaussian process-based fault classification in industrial processes, *Computers & Chemical Engineering*, **170**, 1-11 (2023).

[3] Basha, N., Ibrahim, G., Choudhury, H.A., Challiwala, M.S., Fezai, R., Malluhi, B., Nounou, H., Elbashir, N., Nounou, M., Bayesian-optimized Neural Networks and their application to model gas-to-liquid plants, *Gas Science and Engineering*, **113**, 1-11 (2023).

[4] Malluhi, B., Basha, N., Fezai, R., Ibrahim, G., Choudhury, H.A., Challiwala, M., Nounou, H., Elbashir, N., Nounou, M., Multiscale Bayesian PCA for robust process modeling of a Fischer–Tropsch bench scale process, *Chemometrics and Intelligent Laboratory Systems*, **240**, 2023, 104921.

[5] Fezai, R., Malluhi, B., Basha, N., Ibrahim, G., Choudhury, H.A., Challiwala, M.S., Nounou, H., Elbashir, N., Nounou, M., Bayesian optimization of multiscale kernel principal component analysis and its application to model Gas-to-liquid (GTL) process data, *Energy*, **284**, 2023, 129221.

159b Modeling and Simulation of Blue Ammonia: Challenges and Best Practices for Ammonia – Urea Processes

Fadwa Eljack, Ahmed AlNouss, Fadwa Eljack

Modeling and Simulation of Blue Ammonia: Challenges and Best Practices for Ammonia – Urea Processes

Amzan Alsabri[a], Ahmed AlNouss[b], Fadwa Eljack[a]

[a] Department of Chemical Engineering, College of Engineering, Qatar University, P.O. Box-2713, Qatar

[b] Division of Sustainable Development, College of Science and Engineering, Hamad Bin Khalifa University, Qatar Foundation, Doha, Qatar

*Corresponding Author's E-mail: Fadwa.Eljack@qu.edu.qa

Keywords: Ammonia, Urea, Simulation, Challenges, Best practices

Topic: Computing and Systems Technology Division

Session: 63 Process Modeling and Simulation

Abstract

The ammonia and urea synthesis reactors are the most important units in urea production. Rigorous simulation of the urea reactor is necessary and significant. The modeling and simulation of this system is challenging due to the complex thermodynamic system and chemical reactions.

The market size of nitrogen-based fertilizers has expanded significantly due to the high global demand. At a compound annual growth rate (CAGR) of 5.1%, the market size increased from $87.48 billion in 2022 to $91.97 billion in 2023, and it is expected to reach $111.84 billion in 2027[1]. Urea is the most manufactured type of nitrogen-based fertilizers, where the global capacity is expected to increase by a net of 17 metric tons annually. In 2021, the total production capacity of urea has been estimated at 226 metric tons [2].

Due to the significance of urea production, as well as the availability of modern flow sheeting tools, this study employs basic thermodynamic principles and software engineering to create a tool that can be utilized for modelling key aspects of the urea production processes that are commercially available. While there are several urea production technologies available [3, 4, 5, 6,7], our modelling efforts were focused on the Stamicarbon and Snamprogetti processes because they account for roughly 76% of the global market [8]. In this paper we present a mechanism model using rigorous thermodynamic model, reaction kinetics and fluid dynamics. The urea reactor model is formulated as a set of non-linear equations; it is solved using a stage-by-stage method. The numerical solution of the equations is performed using a conventional iterative method, while the algorithm is simple and robust. Moreover, a literature review was conducted to address the most recent findings in this research scope and compare it with the challenges raised in this study.

References

[1] *Nitrogen Fertilizer Market Demand, Growth Analysis, Outlook 2032*. (n.d.). https://www.thebusinessresearchcompany.com/report/nitrogen-fertilizer-global-market-report

[2] International Fertilizer Industry Association, http://www.fertilizer.org/index.htm

[3] L. W. Codd, 'Chemical Technology: An Encyclopaedic Treatment: The Economic Application of Modern Technological Developments ', Barnes and Noble, 1975

[4] J.A. Kent, 'Riegel's Handbook of Industrial Chemistry', Van Nostrand Reinhold Company Inc, 1983

[5] I. Froment, F. Gilbert, '4th International Symposium on Large Chemical Plants', Antwerp, 1979, Elsevier; Scientific Publishing Company, 1979

[6] R.A. Meyers, 'Handbook of Chemical Production Processes', McGraw-Hill, 1986

[7] 'Production of Urea and Urea Ammonium Nitrate', European Fertilizer Manufacturers' Association, 1995.

[8] Dente, M.; Pierucci, S.; Sogaro, A.; Carloni, G. and Rigolli, E.; 'Simulation Program for Urea Plants', Comput. Chem. Engng., Vol. 21., No. 5, 389-400, 1988

159c Model Development of Pervaporation for Separation of Complex Non-Ideal Mixtures

Andras Jozsef Toth, Huyen Trang Do Thi, Eniko Haaz

Given rising energy costs and environmental constraints, energy conservation remains a top priority for many industries. To increase the profitability of industries and reduce their environmental impact, many methods have been developed to analyze the energy systems of new and existing plants. Pinch analysis is the most commonly used among the various process integration methods used to minimize excessive thermal energy consumption in different industrial processes. The pinch design technology has made it possible to design new plants with optimal energy and capital costs and improve the performance of existing methods. The technique is used worldwide to target various processes' hot and cold energy needs. Based on the pinch analysis, retrofitting the heat exchanger network is one of the promising options for reducing energy consumption, resulting in increased economic and environmental sustainability.

During the work, a process for separating ethanol-water and n-propanol-water mixtures was designed using distillation and hydrophilic pervaporation methods. This method proved effective for the efficient and economical separation of the two azeotropic mixtures. In the case of separating the ethanol-water mixture, it was possible to recover both water and ethanol with a purity of 99% by mass. Even when separating the n-propanol-water mixture, it was possible to recover a product of 99% by weight. However, much larger membranes should have been used, and thus, introducing a second distillation column proved to be more efficient from an economical and quality point of view. Optimization of the heat exchanger system of the planned process is possible based on the conducted pinch analysis and should even be implemented, as it reduces energy consumption and thus results in economic and environmental sustainability.

AUTHOR INDEX